FLORA OF TROPICAL EAST AFRICA

THEACEAE

B. Verdcourt

(East African Herbarium)

Trees or shrubs, usually evergreen. Leaves alternate or spirally arranged, exstipulate, often toothed. Flowers unisexual or hermaphrodite, regular, usually solitary, rarely paniculate, racemose or cymose, frequently large and showy. Sepals 5(–7), free or slightly joined, imbricate. Petals 5(–10), free or slightly joined, imbricate or contorted, spreading except in *Melchiora*. Stamens usually numerous, in several whorls, hypogynous, free, shortly connate or epipetalous. Ovary superior, 1–5-locular ; ovules 1–many in each locule, axile ; styles free or connate.

A family of about 20 genera and over 200 species occurring mostly in east Asia and America. According to a recent investigation made by N. K. B. Robson, *Ficalhoa* appears to be related to *Eurya* Thunb. (*Ternstroemieae, Adinandrinae*) and therefore it is convenient to include it in the *Theaceae* although it has many characters which warrant treating it as a separate family.

This family contains the tea plant (*Camellia sinensis* (L.) O. Ktze. with many varieties), a major crop in East Africa, being cultivated in many forest soil areas with the requisite high rainfall. Since it has on rare occasions occurred as an escape the genus is included in the following key. Several other species of *Camellia* are also cultivated within the area, e.g. *C. sasanqua* Thunb. (Tanganyika, E. Usambara Mts., Amani, *Greenway* 2907, 2972.

Flowers solitary :
- Leaves entire and with inrolled margins, spirally arranged ; tree to 12 m., with solitary, dioecious flowers ; style very short . . . . . . 1. *Ternstroemia*
- Leaves toothed, alternate or spirally arranged ; flowers hermaphrodite :
  - Petals linear-oblong ; corolla appearing tubular ; style very long ; tall trees up to 30 m. . . . 2. *Melchiora*
  - Petals round or ovate ; corolla open, rose-like ; cultivated shrubs or small trees . . . . *Camellia*

Flowers in solitary or paired dichasia . . . . . 3. *Ficalhoa*

Note. *Ternstroemia* and *Melchiora* are more closely related to each other than either is to *Camellia* ; but whilst both the first two have indehiscent fruits, those of the third are dehiscent.

1. TERNSTROEMIA

L. f., Suppl. 39 (1781), *nom. conserv.*

Adinandrella Exell in J.B. 65, Suppl.: 30 (1927)

Evergreen trees. Leaves spirally arranged, simple, subcoriaceous, margins entire. Flowers axillary, hermaphrodite or, more rarely, dioecious (as in the African species). Sepals 5–7. Petals mostly 5, 7–10 in the African

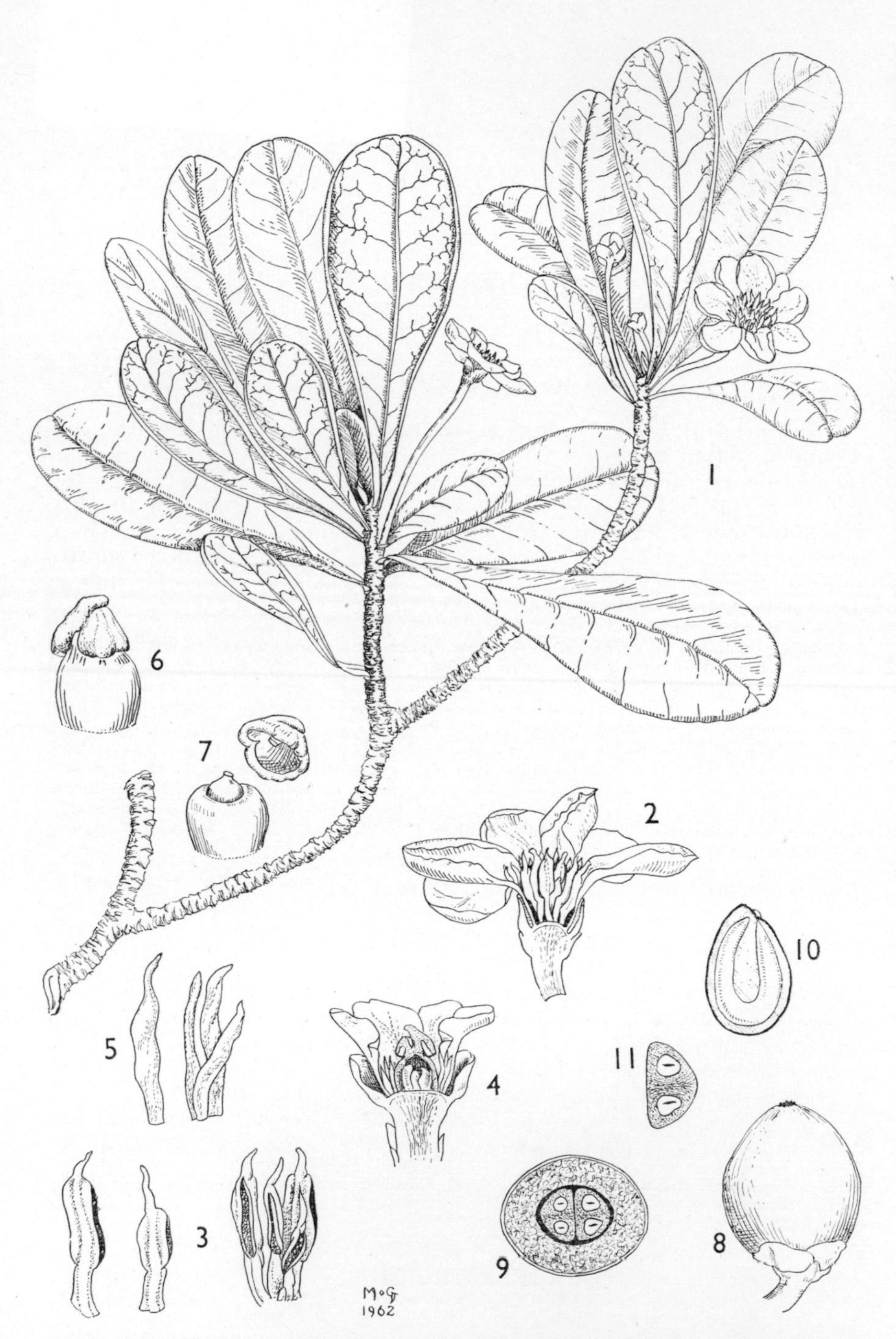

FIG. 1. *TERNSTROEMIA POLYPETALA*—**1,** flowering branch, × 1; **2,** ♂ flower, vertical section, × 2; **3,** stamens, × 4; **4,** ♀ flower, vertical section × 2; **5,** staminodes, × 4; **6,** pistil × 4; **7,** pistil with stigma removed to show style, × 4; **8,** fruit × 2; **9,** fruit, transverse section, × 2; **10,** seed, longitudinal section, × 3, **11,** seed, transverse section, × 3. 1–3 from *Drummond & Hemsley*, 1647; 4–7 from *D.K.S. Grant*, 992; 8–11 from *E. M. Bruce*, 996.

species, opposite the sepals or irregularly disposed. Stamens numerous, 1–many-seriate ; connective of anthers often prolonged beyond the thecae. Ovary (reduced in male flowers of dioecious species) (1–)2–3-locular or falsely 4–6-locular ; ovules (1–)2–20 in each loculus. Style very short with stigmas small, conspicuous or sometimes foliaceous. Fruit indehiscent. Seeds ellipsoid, with a thick testa ; endosperm usually very reduced, sometimes lacking.

The African species belong to the section *Adinandrella* (Exell) Melchior.

T. polypetala *Melchior* in N.B.G.B. 11 : 1095 (1934) ; T.T.C.L. : 606 (1949). Type : Tanganyika, NW. Uluguru Mts., Lupanga Peak, *Schlieben* 3152 (B, holo. †, BM, iso. !)

Glabrous tree (3–)8–12 m. tall. Leaves spirally arranged at ends of shoots, reddish-brown when dry, somewhat leathery, obovate-elliptic, rounded or obtuse at the apex, 4·5–9 cm. long and 1·5–2·5 (–4 fide Melchior) cm. wide ; margins entire, inrolled ; petiole 2–6 mm. long, glandular and flattened, gradually merging with the leaf-blade. Flowers white or yellowish-white, solitary, axillary, dioecious, on rather compressed peduncles 2–3 cm. (–3·5 cm. in fruit) long. Sepals 5, round or ovate, 2·5–4 mm. long and 5 mm. wide. Petals 7–10, obovate, rounded at the apex, narrowed below, 8 mm. long and 6 mm. wide. Stamens about 60 in male flowers, in several whorls, 4–5 mm. long, the anthers much longer than the filaments ; connective produced into an apiculus ; stamens in female flowers fewer and much reduced, in two whorls. Ovary in female flowers ovoid-conic ; stigma conspicuous, cap-like, covering the short style and apex of the ovary. Fruit indehiscent, ovoid, 1·2–1·6 × 0·7–1·0 cm. Seeds 2–4, ellipsoid, 9 × 5·5 mm. Fig. 1.

TANGANYIKA. Morogoro District : NW. Uluguru Mts., Lupanga Peak, 23 Dec. 1933 (fl.), *Michelmore* 853 ! & 11 June 1950 (fr.), *Wigg* 938 ! & S. Uluguru Forest Reserve, Lukwangule Plateau, 17 Mar. 1953 (fl.), *Drummond & Hemsley* 1647 !
DISTR. **T**6 ; known only from the Uluguru Mts.
HAB. Mist-forest and upland rain-forest ; 1800–2300 m.

var. ? ; Brenan, T.T.C.L. : 606 (1949)

Tree 7·5–18 m. tall ; leaves usually less rounded at the apex than in typical material and with rather longer and more distinct petioles ; the peduncles are also shorter, 1·6–2 cm. long.

TANGANYIKA. Iringa District : Dabaga Forest Reserve, Oct. 1958, *MacHattie* H337/58 ! & 18 Oct. 1937, *Pitt-Schenkel* 577 ! & Mufu Ukwama, 6 Sept. 1958, *Ede* 23 !
DISTR. **T**7 ; not known elsewhere
HAB. Camphor-bamboo-forest, 1500 m.

NOTE. The small differences mentioned above may not be significant. More material from other areas is needed.

2. MELCHIORA

Kobuski in Journ. Arn. Arb. 37 : 154 (1956)

Adinandra Jack sect. *Eleutherandra* Szysz. in E. & P. Pf. 3 (6) : 189 (1893)

Adinandropsis Pitt-Schenkel in Journ. Ecol. 26 : 80 (1938), *nom. nud.*

Evergreen trees. Leaves alternate, simple, papery or subcoriaceous ; margins closely serrate ; lateral nerves numerous. Flowers hermaphrodite, solitary in the leaf-axils ; bracts 2, persistent. Sepals 5, imbricate, concave, persistent, very unequal, the outer ones the shorter. Petals 5, free, connivent in a tube. Stamens uniseriate, 15–35, free amongst themselves but adnate to the corolla at the base. Ovary 4–5-locular, many-ovuled. Style long, filiform ; stigma slightly 5-lobed. Fruit indehiscent, enclosed in the persistent sepals. Seeds many in each locule, small, reniform, shining.

M. schliebenii (*Melchior*) *Kobuski* in Journ. Arn. Arb. 37 : 156 (1956) & 38 : 202, plate 2 (1957). Type : Tanganyika, NW. Uluguru Mts., Lupanga Mt., *Schlieben* 3175 (B, holo.†, A, BM !, BR !, K !, iso.)

Tree 6–40 m. tall, with straight clean trunk ; probably all the varieties have a rough reticulate bark. Leaves subcoriaceous, oblanceolate or elliptic-oblanceolate to broadly elliptic, finely glandular-serrate, cuneate at the base, ± abruptly shortly acuminate at the apex, glabrous above, glabrous to densely velvety below, 5–15·5(–18) cm. long and 1·5–6·7 cm. wide ; petiole 5–6(–10) mm. long. Flowers on peduncles (1–)2–3·5 cm. long ; bracts 2, opposite, immediately below the calyx, rounded or subacute, unequal, 2–9 mm. long and wide. Sepals reddish or brownish, ovate, ± acute or subrotund at apex, 1–3 cm. long and 1·2–1·6 cm. wide, glabrous or pubescent outside (outer always glabrous), finely sericeous within. Petals mostly orange or orange-red with paler tips, linear-oblong, 3·2–5·0 cm., mostly 4·3–4·5 cm. long and 5–7 mm. wide, obtuse or acute at the apex, subentire or slightly denticulate, narrowed below. Stamens 1·1–2·3 cm. long. Ovary conical, 5–10 mm. long, glabrous or hairy. Style filiform, 2·4–5 cm. long. Fruit ovoid-conic, 2–3 cm. long and 1 cm. diameter. Seeds numerous, 1·0–1·5 mm. in diameter.

KEY TO INTRASPECIFIC VARIANTS

Ovary glabrous . . . . var. **glabra**
Ovary pilose :
  Leaves pubescent to velutinous below . . . var. **greenwayi**
  Leaves entirely glabrous :
    Inner sepals more or less pubescent outside . . var. **schliebenii**
    Sepals glabrous outside. . . . . . . var. **intermedia**

var. **schliebenii**

Tree 10–30 m. tall. Leaves glabrous. Calyx ± 1·6 cm. long. Inner sepals silky pubescent outside, particularly when young. Corolla orange and red or orange with yellow tips to the lobes.

TANGANYIKA. Morogoro District : Uluguru Mts., Bondwa Peak, 23 Mar. 1953, *Drummond & Hemsley* 1766 ! & Lupanga Mt., 28 Dec. 1932, *Schlieben* 3175 !
DISTR. **T6** ; known only from the Uluguru Mts.
HAB. Slopes of upland rain- and mist-forest, a dominant species sometimes forming 70% of the tree-cover ; 1400–2000 m.

SYN. *Adinandra schliebenii* Melchior in N.B.G.B. 11 : 1076, 1097 (1934) ; Kobuski in Journ. Arn. Arb. 28 : 95 (1947) ; T.T.C.L. : 605 (1949) ; Verdc. in K.B. 10 : 608 (1956).

var. **intermedia** (*Boutique & Troupin*) *Kobuski* in Journ. Arn. Arb. 38 : 203, plate 3 (1957). Type : Congo Republic, District of Lakes Edward and Kivu, Mayamoto, *Michelson* 742 (BR, holo !, A, EA !, iso.)

Tree 21–30 m. tall ; bark rough, brown and reticulate, similar to that of *Ocotea* ; trunks solitary or 2–3 together producing stilt roots about 1 m. above ground. Leaves quite glabrous. Calyx 2·3–2·5 cm. long ; sepals glabrous outside, very finely silky-pubescent inside. Corolla orange-yellow, brick-red or orange-red with yellow base.

UGANDA. Kigezi District : Impenetrable Forest, Sept. 1936, *Eggeling* 3274 ! & Oct. 1940, *Eggeling* 4167
TANGANYIKA. Lushoto District : W. Usambara Mts., Shume, 25 Aug. 1952, *Greenway & Parry* 8740 ! & Magamba Forest, Aug. 1934, *Pitt-Schenkel* 376 !
DISTR. **U2** ; **T3** ; eastern Congo Republic, Ruanda-Urundi
HAB. Upland rain-forest ; 1900–2350 m.

SYN. *Adinandra intermedia* Boutique & Troupin in B.J.B.B. 20 : 62, f. 8–10 (1950) [*A. schliebenii* sensu Dale, I.T.U., ed. 2, 424 (1952), *non* Melchior]
A. schliebenii Melchior var. *intermedia* (Boutique & Troupin) Verdc. in K.B. 10 : 608 (1956)

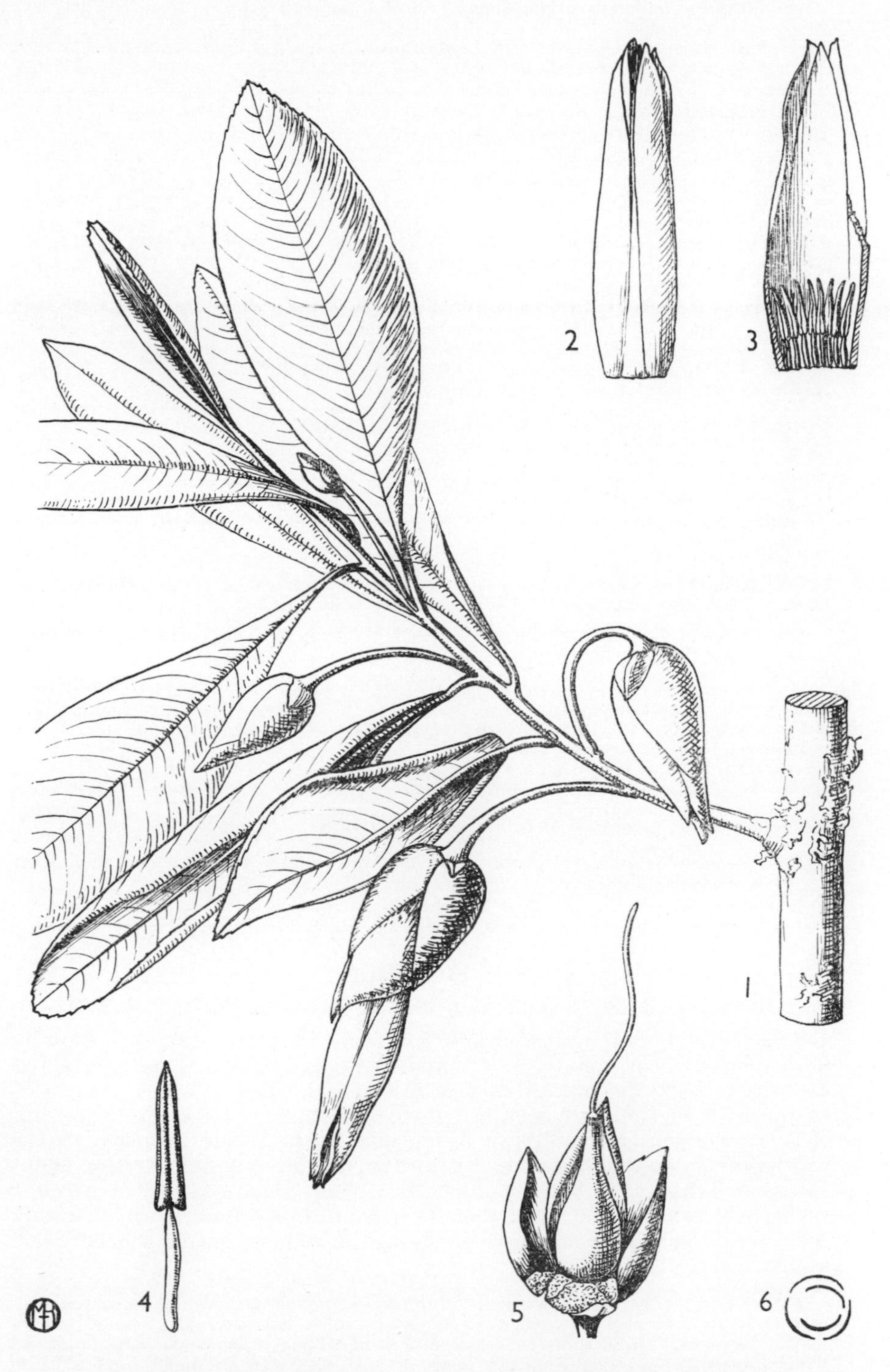

FIG. 2. *MELCHIORA SCHLIEBENII* var. *GLABRA*, from *Drummond & Hemsley* 2536—**1,** flowering branch, × 1 ; **2,** connivent petals, × 1 ; **3,** 3 connivent petals, from within, showing stamens, × 1 ; **4,** stamen, × 3 ; **5,** pistil (and 3 sepals), × 1 ; **6,** diagram to show aestivation of sepals.

Melchiora intermedia (Boutique & Troupin) Kobuski in Journ. Arn. Arb. 37 : 157 (1956)
" *Adinandropsis* sp. nov." Pitt-Schenkel in Journ. Ecol. 26 : 80 (1938)
" *Adinandra* sp. near *schliebenii* " Brenan, T.T.C.L. : 605–6 (1949)

var. **greenwayi** (*Verdc.*) *Kobuski* in Journ. Arn. Arb. 38 : 204, plate 4 (1957). Type : Tanganyika, Pare District, Mtonto, *Greenway* 6556 (K, holo. !, A, EA, iso. !)

Tree 15–24 m. tall ; bark reddish-brown, fissured and reticulate. Leaves when mature, either densely covered with orange-brown or creamy-brown tomentum beneath or at least with a band of tomentum on each side of the costa. Calyx 2·0–2·6 cm. long. Sepals glabrous outside, or inner ones pubescent near tips only. Corolla creamy-brown, with petals tipped with green. Ovary densely hairy, the hairs extending further up the style than in the other varieties with hairy ovaries.

TANGANYIKA. Lushoto District : W. Usambara Mts., Shume-Magamba Forest, 15 July 1954, *Hughes* 203 ! ; Pare District : Pare Mts., Mtonto, 5 July 1942, *Greenway* 6556 !
DISTR. **T3** ; not known elsewhere
HAB. Upland rain-forest ; ± 1920–1950 m.

SYN. *Adinandra greenwayi* Verdc. in K.B. 8 : 84–5 (1953)
A. schliebenii Melchior var. *greenwayi* (Verdc.) Verdc. in K. B. 10 : 608 (1956)

NOTE. In K.B. 10 : 608 (1956) I cited *Greenway & Parry* 8740 under var. *greenwayi*. On further examination I refer it to var. *intermedia*, although there are traces of tomentum near the costa beneath. Var. *greenwayi* is in effect merely a hairy form of var. *intermedia*.

var. **glabra** (*Verdc.*) *Kobuski* in Journ. Arn. Arb. 38 : 203 (1957). Type : Tanganyika, Lushoto District, Shagayu Peak, *Procter* 183 (EA, holo. !, K, iso. !)

Tree 6–12 m. tall. Leaves glabrous. Calyx 2·0–2·7 cm. long. Sepals glabrous outside but retaining the very fine sericeous tomentum within. Corolla orange-red or orange with yellow or greenish-yellow tips. Fig. 2, p. 5.

TANGANYIKA. Lushoto District : W. Usambara Mts., Shagai Forest, Shagayu Peak, May 1953, *Procter* 183 ! & 15 May 1953, *Drummond & Hemsley* 2536, & Sungwi Forest Reserve, Aug. 1955, *Semsei* 2303 !
DISTR. **T3** ; known only from the W. Usambara Mts.
HAB. Drier parts of upland rain-forest and on exposed tops of ridges ; 2220–2230 m.

SYN. *Adinandra schliebenii* Melchior var. *glabra* Verdc. in K.B. 10 : 608 (1956)

NOTE. Possibly a form developed in windy exposed situations but probably more than just an ecological form.

3. FICALHOA

Hiern in J.B. 36 : 329 (1898) ; Robson in Fl. Zamb. 1 : 405 (1961)

Tree with terete branches and alternate serrulate leaves. Flowers hermaphrodite, in solitary or paired, axillary, few- or many-flowered dichasia ; bracteoles 0 or 2. Sepals 5, rarely 6. Petals 5, rarely 6, alternating with the sepals, connate in the lower third to form a suburceolate corolla. Stamens in bundles of 3 opposite to the sepals, attached to the corolla-tube ; anthers opening by means of apical pores. Ovary hemispherical, obtusely 5–6-lobed, 5–6-locular ; ovules small and very numerous ; styles 5–6, connate for 2/3 of their length or almost free. Fruit a woody 5–6-valved hemispherical capsule. Seeds numerous, small, winged, with scanty endosperm.

A monotypic genus, placed originally in the *Ericaceae* but, although perhaps distantly related to that family, apparently most closely allied to *Eurya* and related genera in the *Theaceae*. From these genera it differs in having a loculicidal capsule, small winged seeds with a straight embryo borne on a swollen placenta and a copious white latex. Cymose inflorescences are very unusual in the *Theaceae* but occur in some species of *Eurya*.

FIG. 3. *FICALHOA LAURIFOLIA*—**1,** flowering branch, × 2/3 ; **2,** flower, × 8 ; **3,** corolla opened to show stamens, × 8 ; **4,** ovary, × 8 ; **5,** dehisced capsule, showing sterile ovules on the 5 swollen placentae, × 8 ; **6,** seed, × 24. 1–4, from *Angus* 841 ; 5, from *Osmaston* 2060 ; 6, from *Davies* 297. Reproduced by permission of the Editors of " Flora Zambesiaca ".

F. laurifolia *Hiern* in J.B. 36 : 329, t. 390 (1898) and in Cat. Afr. Pl. Welw. 1 : 633 (1898) ; V.E. 1 (2) : 670, f. 578 (1910) ; Z.A.E. 2 : 509 (1913) ; F.P.N.A. 2 : 16 (1947) ; T.T.C.L. 193 (1949) ; I.T.U., ed. 2, 110 (1952) ; Robson in Fl. Zamb. 1 : 405 (1961). Types : Angola, Huila, near Lopollo, *Welwitsch* 4808 (BM, syn. !, K, isosyn. !) & between Mumpulla and Nene, *Welwitsch* 4809 (BM, syn. !)

Evergreen tree (2–)6–24(–36) m. tall, much branched, with rough fissured bark and a copious white latex. Branches at first sparsely pilose, eventually glabrous or in some specimens densely pilose. Leaves somewhat leathery, oblong or lanceolate, acuminate at the apex, cuneate or rounded at the base, (5–)7–12·4 cm. long and (1·8–)2–4·3 cm. wide, with margins bluntly serrulate, glabrous or sparsely pilose ; petiole 3–9 mm. long. Flowers white, yellowish or greenish, in solitary or paired cymes ; peduncle 2–6 mm. long, pubescent or pilose ; pedicels short, pubescent. Sepals rounded, sparsely to densely pubescent, about 1 mm. long. Petals oblong, rounded 3 mm. long. Ovary densely adpressed-pilose ; styles pilose at the base. Capsule about 3 mm. in diameter, with widely spreading valves. Seeds about 0·5 mm. long. Fig. 3, p. 7.

UGANDA. Toro District : Bwamba Pass, July 1940, *Eggeling* 4029 ! ; Kigezi District : Impenetrable Forest, Sept. 1936, *Eggeling* 3259 !

TANGANYIKA. Mpanda District : Kungwe Mt., Selimweguru, 24 July 1959, *Newbould & Harley* 4630 ; Njombe, Aug. 1933, *Greenway* 3508 ! ; Songea District : Matengo Hills, Luwiri-Kitesa Forest, 25 Oct. 1956, *Semsei* 2548 !

DISTR. **U**2 ; **T**3, 4, 6–8 ; Congo Republic, Ruanda-Urundi, Nyasaland, Northern Rhodesia and Angola

HAB. Upland rain-forest, riverine forest and bushland, streamsides and cultivated ground in areas cleared of forest ; 1350–2400 m.

NOTE. Some Tanganyika material (e.g. *Carmichael* 98 from Iringa District, Msombe Hill) has very pilose branches and sepals.

INDEX TO THEACEAE